Bertha Bridges

The Cow on the Farm

Bertha Bridges

The Cow on the Farm

ISBN/EAN: 9783337183066

Printed in Europe, USA, Canada, Australia, Japan

Cover: Foto ©berggeist007 / pixelio.de

More available books at **www.hansebooks.com**

THE COW ON THE FARM

BY

Bertha Bridges

✳

CINCINNATI

The Robert Clarke Company Press

1896

PREFACE.

This little book has been written by a woman of nearly twenty years of actual experience in the business of making butter. It is intended for the women on the farm who may have a desire to perfect themselves in the art of making and preserving butter. The writer has taken great pains to give the best information. If our cows are properly taken care of and the butter is made without violating certain laws, our butter will be as perfect in November as it is in May, and there is no reason why any one can not make good, palatable butter the year round. The writer hopes that those looking for information may find in this little book an ever-present help.

THE AUTHOR.

THE COW ON THE FARM.

FEED AND CARE OF THE COW.

Many farmers who in nearly all of their farm duties are neat and careful, are careless in the management of their cows. They do not consider what a difference it would make if their cows were properly cared for. Instead of being housed and fed in a clean stable, they live in the barnyard or strawstack. Their coats are filthy, every rib can be counted, and a man can hang his hat on their hip bones. And these poor, neglected creatures are expected to give good, wholesome milk! Is it a wonder that some butter is not worth ten cents a pound?

This, however, is not the state of affairs every-where, and I am glad to say it is the exception and not the rule. Most farmers have long since come to find out that our farms without good, well-kept cows, and a flock of equally well-kept hens, would hardly be farms or *homes* either. In summer, it is, of course, unnecessary to stable cows, as they do much better in the pasture or lot at night. But as soon as the weather gets stormy, they should be housed in a warm, clean place, being turned out, however, every day, even during cold and snowy weather, as they need exercise to keep them well and thriving. Feeding also should be judiciously done. Rotten corn and moldy fodder will not furnish good milk and make good-flavored butter. In winter, when there is no pasture, it is a good plan to feed some mill feed, such as bran or what is called shipp-stuff. Yellow corn, ground cob and all, also makes good winter feed. Good, clean clover hay is very good to give the butter a fine color, so much desired in winter. One very cold winter I fed yellow corn and sugar beets with good fodder for forage. My cows were fat and sleek and turned off a wonderful amount of butter of good color and delicious flavor.

In summer, when pastures are very short and the cows eat ragweed and also horseweed, the butter will get very soft and oily, and there seems to be no grain in it. When this is the case, it is a good plan to feed a little green corn, stalks, blades, corn, and all. Where one has sugar corn, the suckers which are usually pulled and left in the field will do. The farmer who saves these and feeds them to his cows may consider the time that it takes to do so well spent. This feeding will not only improve the flow of the milk and quality of the butter just at a time when there is a great demand for the latter, but it will also keep the cows quiet, and will often prevent their becoming troublesome where fences are poor. Cows should always have plenty of *good, pure* water, and have salt once or twice a week. When the cows do not have salt enough the butter will be hard to churn.

In the summer, when the cows are running in a large pasture, it is well to give them a bit of something to eat at night, so as to make them come home at milking time. There are many things about a farm that will answer this purpose. A few handfuls of bran, in the spring, a little green clover, a few nubbins of sugar corn with the husks from the corn used for the dinner table, or a few dropped or specked apples, will make a cow come half a mile, if they are seasoned with a few kind words and a little petting. And this labor-saving plan will pay well on any farm.

At times when they are housed they should also be fed at milking time, as this puts them in a good humor and makes them give their milk down. The droppings should be removed before milking, and the milking should be done in an even, steady manner. The milk should froth and foam in the bucket. Also, the cows should always be milked at the same time of day, and not at five o'clock to-day and seven o'clock to-morrow. I do not mean by this that we should milk our cows at the same hour the year round; but the change must be so gradual that it will not injure the cows. Every drop of milk should be taken, for when this is not done the cow will go dry before it is time for her to do so. The regular time for a cow to go dry is six weeks. Sometimes it is necessary to milk the cow a few days before she has her calf. Cows should always be noticed at this time; and if the udder appears inflamed, they should be attended to, as neglect may give trouble after they are fresh. Should the udder be caked and inflamed after they are fresh, hog's lard, heated as

hot as one's hand will bear, is good to
bathe with; also to bathe with the milk
which is milked from the cow is good to
soften the udder and to remove inflam-
mation. In all cases, however, the pre-
ventative is better than the cure, and a
little care taken at the proper time will
often save much trouble.

Cows Should Never be Forced.

By this we mean that they should not
have more than regular rations. Many
people think that by giving their cows
extra feed before coming fresh that they
will do better afterward. This treatment
is unnatural and will often end in the
loss of a good cow. A cow at the time
of coming fresh should have good, warm,
soft food. This is made by taking say
a wooden bucket half full of mill feed;
any kind of good mill feed will do.
Give with this *for the first feed* a common
sized fire shovel nearly full of good wood
ashes and a small handful of salt, and

fill the bucket with warm water. This
will, if the cow is healthy, be all that is
required to accomplish whethering. For
the next few feeds it is well to add a
handful or two of oil meal. The cow
should be fed moderately for a few days,
and should have *warm drinks only.* In
very cold weather it is at all times best
to warm the water for the cows to
drink.

RAISING THE CALF.

Raising a young calf is often a difficult
task, and unless they are attended to
very carefully will not do well and often
get sick and die. The milk fed to a calf
should never be more than milk warm
(blood heat); next it should always be
perfectly sweet. If the milk fed to a calf
is too warm there is danger of killing
them almost instantly, and if it is in the
least off or sour it will give them dysen-
tery; in the latter case they can usually
be saved, so long as they will eat, by

giving them a mess or two of boiled milk. After boiling the milk let it cool until it is milk warm and give them about half rations for a few feeds, and as soon as they are well give them the same quantity as before, increasing however gradually. For the first week or ten days the calf should have fresh milk warm from the cow. After this skimmed milk will do. Where the milk is needed for the family, a little gruel can be added after the calf is six or eight weeks old. In giving gruel, begin with a very little at first; use for the first feed a small handful of meal and boil this well. If it takes this and there are no bad effects, then give this for a few feeds, and if it continues well increase the portions gradually until you can give it half gruel and half milk. This drink should be given until the calf is at least four months old. After this the gruel without the milk will do.

When depending on our own resources,

the proper way to feed a calf is to take
a vessel part full of the warm sweet milk,
put the palm of your hand to the calf's
nose and into the milk, slipping your
finger under its upper lip, and if the calf
has not been sucking the cow it will
learn to eat very quickly, and will after
being fed a few times take the milk without any trouble. A calf should have its
food regularly, and it should, if this is
possible, be prepared by one person only ;
by doing this there is less danger of over-
feeding or of giving it something it ought
not to have. The little fellows are apt
to eat very fast sometimes and must be
held back, as it is not good for them to
drink the milk too fast. The right person may raise a number of calves suc-
cessfully, and yet, if they should be com-
pelled for any reason to give the care of
one up to some one else, for a single feed,
may lose it. They must have good, nat-
ural, common sense treatment, and when
getting this will usually thrive and live.

THE STABLE AND BARN.

A bank stable is very practicable for cows where one can have it, and a dirt floor is best if the stable is well drained. A stable built with the north and west side a good stone wall will be very warm in winter. Each cow should have about five feet in the width and at least ten in the length of her stall, besides manger and feed box. Where this can be done, it is well to have one door of the stable open into the pasture, as this saves much trouble in handling the cows as then they need only be untied and can go direct into the pasture. And when put up the door only need to be opened for them to walk to their places. I have my stable arranged in this way and find that it is very convenient, and nearly as warm as a cellar. For this stable, see Plan No. 1. It is built 18 x 20 feet, and is one and one-half story. Where there is no other barn, however, to store feed

Bank

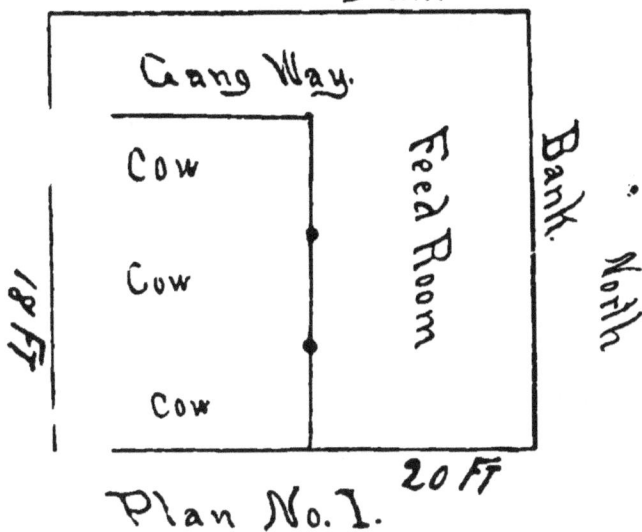

Plan No. 1.

for forage, the upper story can be built higher.

Where a larger barn is wanted, Plan No. 2 will be suitable for eight cows. This barn is 20 x 30 feet, and is three stories. The first floor is for the cows; the second is made with bins to store grain, mill-feed, beets, or other tubers intended for feeding. This floor is to come out level with the ground at the back, so one can drive in. The third floor is to be used for hay, and is connected by a chute to the first floor to drop the hay down; this chute to come out at the end of the passage between the cows. The cows can be fastened with stanchions, but do not need partitions between them.

BE KIND TO THE COW.

While we should be kind to all animals it is most necessary in the treatment of our milch-cows. Let them know that you are master once for all. Never tolerate any misdemeanor. A cow must

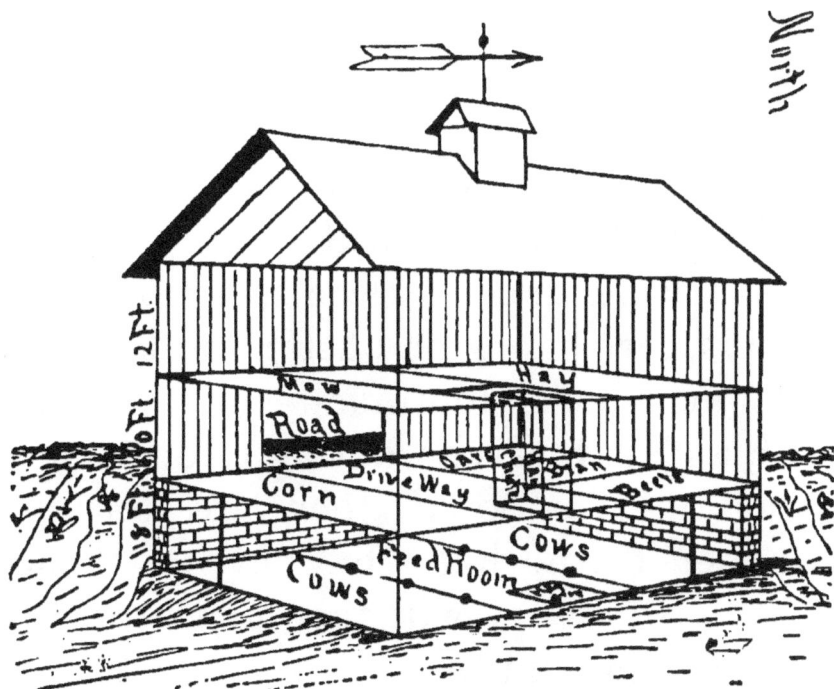

North

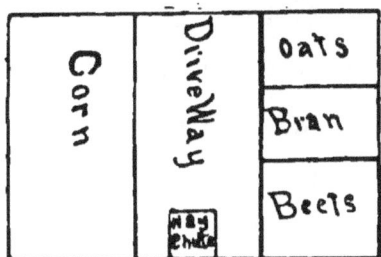

Second Floor.

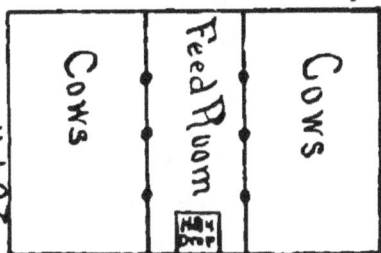

First Floor

North

Plan No. 2.
Interior View.

know that she dare not raise her foot to kick the bucket over, or to strike at the one that is milking. They *must be taught respect*, but they can also be taught to love us in their own brute way. A cow with a kind and gentle disposition will always give more and better milk than the one who is continually on the war path and is forever wanting to do something mean. So if they are unruly conquer them, but be kind to them afterward. They will then first learn to respect you, and as soon as they know that there is nothing to fear will learn to love you as well as a faithful dog. Remember that any thing worth doing at all is worth doing right. So let us persevere, and after a while if we try and work hard enough we will get to the top of the ladder. We must not, however, go ahead as one blindfolded, but work and watch our path as we are going onward, so we may *keep on the right track*, and this not just in one branch of our work but in every thing.

MILK AND BUTTER.

Now that the cow has been properly cared for we will expect her to give good milk, and then we can be expected to make good butter. Good Butter. How much do these two simple words signify to the farmer's wife. And how much more do they imply to the thousands consuming strong butter or tasteless oleo. Now, making good butter is not an easy task, though the feeding and care of cows were all done right. In butter making, as well as in every thing else, we should make ourselves acquainted with the best methods, so as to get the best returns for our labor.

CLEANLINESS IS THE FIRST CONDITION ESSENTIAL TO SUCCESS.

The place where the milk is kept should be neat and clean and the *air pure.* If a cellar, it should be white washed every spring and kept perfectly clean. Such things as barrels of vinegar and vegetables have no place in a cellar where milk is kept. A vessel with air-slaked lime and a bucket of pine-tar (the latter should be stirred once a day) will help to keep the air pure and sweet. We all know that milk takes up any impurity out of the air very quickly. Any one that cares to take notice has observed that if we keep milk in our kitchens in winter and then cook vegetables, such as turnips, cabbage or onions, indeed, any thing which smells strong while cooking, this will make the milk taste, and if repeated often the butter will be strong. Indeed, if we would have our butter to be first class every thing about it should be so.

The cupboards, tables, and benches should be clear of mold and should be scrubbed white and clean. Any milk that is spilled should be wiped up im-

mediately. Next to this all milk vessels should be used for milk only, and
should be washed by themselves in *clean
very hot* water, and *without the use of soap.*
After being washed clean and wiped dry,
all stone ware used for milk should be
heated quite hot.   One can turn them on
the stove, or set them in the oven, and
when one lot gets hot take them away
and heat another.   In summer when we
do not have much fire, pile the pans up
in the kitchen and heat them whenever
there happens to be a fire in the stove.
This heating of the milk vessels is indis-
pensable where the butter is intended
for packing.   But should be done at all
times.   In winter as well as in summer.
It is a bad practice to put stone pans in
the sun to sweeten, for no matter how
hot the sun may be, it is *not hot enough
to destroy* the milk which may be left in
the pores of the pans.   It sours and
rots it, and the pans are only worse
than before.   On the other hand, heat-

ing by the fire burns up whatever milk may be left in the pores of the pans, and they become sweet, and then putting them in the sun will not hurt them. Where the pans have been in the sun all day, it is a good idea to fill them with cold water for fifteen or twenty minutes before straining the milk into them. The milk should be strained and put away as soon as brought in, as much of the cream is lost by letting it set in the buckets. Never cover the milk while the cream is raising if you can help it. But if for any reason it becomes necessary then avoid mold on the covers. And when scrubbing, never use soap. Use very clean hot water and let them get perfectly dry before using again. Stone pans are better to strain in than tin for this reason: The milk will eat the tin, and while there may be only a small quantity in each lot, yet in the course of time this will be injurious to health. The stone pans may be more difficult to handle, still I think them preferable.

WHEN AND HOW TO SKIM THE MILK.

In skimming milk, run your finger around the pan, so as to loosen the cream, and with a tin ladle lift it off. Some one once said while watching me at skimming : " How close you are ; you want every bit of cream." And they were right in one point, for I do want every bit of cream, and every one making butter should do the same, for the cream left on the milk or side of pan would only be wasted, and that would be against the principles of economy.

Always aim to have the cream off before the milk is thick. It may be sour, and I think I can make fully a third more butter when the milk can set quietly until it is first turning sour, and then skim. Instead of taking the cream while the milk is still sweet. In summer usually skim thirty-six hours after milking, unless, as sometimes is the case, the weather is very sultry, and then the milk will sour in twenty-four hours.

NEVER *under any circumstances, winter or summer, should milk set longer than forty-eight hours before being skimmed.* (I wish I could sufficiently impress this on the mind of every one trying to make good butter. For I firmly believe that nine out of every ten pounds of bad or indifferent butter is so from this cause.)

Some of our best housekeepers who are among the first in all the other branches of housekeeping have cheesy butter in winter. Their bread is the finest; their cakes and pies are the most delicious. Their linen is the whitest, and, indeed, every thing about the house is in harmony. They do, indeed,

> Guide the house with prudent care,
> With judgment wise to spend and spare,
> And make their husbands bless the day
> They gave their liberty away.

With these women it is neither carelessness nor neglect that makes their butter bad. Many times the food is blamed, and I have often heard the remark made that some sort of winter food would make the butter strong. Now I do not deny that there is food that does make the butter bad. But in many cases where the food is blamed, the butter is bad from a different cause altogether. Let every woman that wishes to know try for herself; it is very easily done. For one week skim the milk at forty-eight hours after milking, and the next week let it set any length of time that may suit, but let it set longer than forty-eight hours and I will tell her right now that she will find that her butter will be cheesy and strong just in proportion to the length of time that the milk has been setting. Often a little thin scum of cream will raise on the milk after it has been skimmed, and this will make many people think that they are

losing by skimming the milk too soon.
If those afraid of this will give it a trial,
they will find that by skimming their
milk at forty-eight hours they will
have just as much if not more butter
than they would have if they would let
it set longer, so they will find that they
will lose nothing in weight and will gain
double in quality. The cream raises
best in a moderately warm place, say a
temperature of not less than 50° nor more
than 60°. Should the milk, however, be
still sweet after setting the proper length
of time, it must nevertheless be skimmed,
as letting it set *longer will make the butter
strong.*

Bitter Butter in Winter.

Many people complain of having bitter
butter in winter. Where this is the
case, it is nearly always caused by letting
the milk set too long before skimming
or by not souring the cream after it is
skimmed. I have found it best to sour

the cream as soon as I begin saving.
Say the first day I take the cream jar
and set it by the stove. Keep turning
until the cream is well warmed through.
Then keep it so for a few hours and next
morning the cream will be sour. Now
keep on skimming into this and the rest
will sour without any trouble. The
cream should be stirred every time new
cream is put into the jar. I will also say
here that cream properly ripened will
turn off nearly a third more and better
flavored butter than it will when it is
churned sweet. I never let milk freeze
if I can help it, as it makes skimming a
tedious task. Now there is one more
difficulty which I will mention here. In
summer during excessive heat the milk
will sometimes for weeks have whey on
the top almost as soon as it turns sour.
Many people think that this is caused
by the cows eating ragweed and horseweed. But I have found it to be the
case when the cows were on good blue

grass and when there was not the least taint of ragweed in the milk. The only thing that I have found to help here is to take the milk as soon as strained and set it on the stove in the pans that it is to be left in. Leave on the fire long enough for the scum to raise. Keep at this heat for half an hour or so, but do *not let it boil.* This heating will keep the milk sweet longer than it would keep by boiling it. The cream will raise nicely; the butter will be firm and grainy and will where there is trouble in churning come sooner. Often, when the cows are nearly dry, there will be trouble in gathering the butter. Where this is the case, if the heating of the milk does not help, a little salt added to the cream each time new cream is put in the jar, or if this has not been done, putting the salt in the churn will often gather the butter very quickly.

The Churn and Churning.

Now the next thing is the churn and churning. As for the churn, I can say very little in favor of patent churns, as far as my experience goes, and am still old fashioned enough to like the "Old Dash" churn best, and can say much in its favor. In the first place it is cheap and very durable; next, it is easily kept clean; and if the cream is good and has the right temperature, we can churn our butter in twenty minutes. The butter is easily taken out, and when all is done as it ought to be, there is not a bit of waste about it.

Churning in Winter.

In winter the churn should always be scalded as well before as after churning. In the summer do the same, but scald your churn in the evening before you intend churning the next morning. The tray and butter ladle should be well soaked before taking the butter up, as this makes them cool and keeps the butter from sticking. The churn should be cleansed as soon as the butter is salted and set away for the salt to melt. Never let the buttermilk set in the churn. To begin churning: After you have scalded the churn, wet well with the water and empty. Then let it set a few moments to cool, and as soon as it is done steaming, it is ready for the cream. The cream should be heated from 62° to 64°. With a little practice a person can soon tell just how warm it ought to be. After the cream is in the churn hold the thermometer in a few moments. If it is not warm enough, add boiling water enough to make it so, and if too warm, add cold water. Adding the boiling water *before beginning* to churn will never scald your butter so long as you only have the cream at 64° when you begin to churn. It is only by adding hot water after the churning is part done that it will scald the butter.

BUTTER SHOULD NEVER BE MUSHY.

Summer before last I had an experience with mushy butter which I will give here as an example. One very hot day one of my neighbors came to me, saying: "Oh, Mrs. B., can you come down to my wife a little bit. She is churning and she can do nothing with her butter." They were young housekeepers. I went down. It was near noon, and here was the woman in the cellar. She had a tub half full of water and milk, and a churn full of the same mixture with three or four pounds of butter that looked like pancake batter floating on it. "Oh," she said, "I can't get the milk out of this stuff. I think it must be our cow that is not good, and I can't make butter out of her cream. I have been trying and trying, and it is always as you see it here." I told her I thought if she would churn early in the morning and out of doors, in place of in the cellar, she would find a difference. I asked her to give me her tray and butter ladle, and I would see what I could do for her. She got them. They were wet, but as soon as I touched the butter it all stuck fast. I cleaned off the butter and asked her to give me some coarse salt. I then rubbed ladle and tray well with the salt, and after rinsing well took up the butter. The mess was so sloppy nothing could be done with it but to salt it, or, rather, to stir the salt into it. I then put it in a crock and told her to let it set in the cellar on the ground, well covered, until morning, when it might be ready to work. And, indeed, after turning out and working next morning, the butter looked respectable. False pride had kept this woman from telling her trouble, until, at last, she found she could not help herself, and then she came to me. I afterward gave her some good practical advice, and she had no more trouble with her cow.

CHURNING IN SUMMER.

I think it far easier for those of little experience to make good butter in winter than in summer. But those who know their business can make good, solid butter even in the hot days of July and August, and do so without the use of ice. And let me say here that solid butter made without ice will stand the heat much better than butter made where ice has been used. As soon as warm weather comes, the churning should always be done very early in the morning, in a place that is cool and shady, but free from a breeze and flies. Three times a week is often enough in summer during hot weather, and as soon as it gets a little cooler twice will do. As I have said before, do not neglect to scald your churn. Do this in the evening before you intend churning next morning. After being scalded, empty and let cool. Then set a bucket of cold water in it over night. In order not to have an extra fire, it is well to do this while getting supper.

When your churning is done and the butter which may be on the side and lid washed down carefully, take your tray and ladle (which should have been well soaked) and take your butter out. Then drain the milk off carefully, spread your butter in the tray, and it is ready for salting.

Salting and Working Butter.

For every six pounds of butter, a tea-cup of *fine table salt* and half a teacup of *granulated* sugar. This method of salting will be new to nearly every one, yet I feel confident that if they once try it they will never give it up. I have been using sugar in my butter for a number of years, but only within the last two years have I used it just in this propor-

tion. I find that it makes the butter
much more solid, and when it has been .
made just as it ought to be made, it will
keep perfect from April until Christmas
and no telling how much longer.

After the butter has been salted and
the salt and sugar simply worked through
enough to mix it well, take the tray and
set it away long enough for the salt to
melt. In warm weather and when the
butter is soft, let it set in a cool place
until it begins to get firm and grainy,
and then work. In working butter, it
should be pressed only. If the ladle is
rubbed over the mass instead of pressing
it, it breaks the grain and makes the
butter look oily. To those who have
never seen butter worked I will say:
Take your butter and with the ladle
roll it over to one side of the tray; then
with the ladle cut off a piece and put on
the opposite side, pressing it down solid;
now take one piece after the other and
press in the same manner, until you have

it all over; drain off what milk you have worked out; turn your butter back and work as before; repeat until the liquid pressed out looks clear, when your butter is done. Never wash butter. While there is still much that could be said about butter making, I think I have said enough to make things plain. I have tried at least to explain the four *fundamental principles*—the first, cleanliness; the second, to skim the milk at the proper time and in the right manner; the third, to do your churning right; and the fourth, to work the butter properly. There is no way of getting around either of these principles. We can not be dilatory or neglectful and meet with success.

EXPERIENCE MAKES WISE.

In the spring of 1895, when butter took such a drop in the market, I was making at the rate of one hundred pounds a month from three cows. I had between fifty and sixty pounds of it engaged to regular customers and the rest I let my storekeeper have. He had been handling my butter for years and had always given me twenty-five cents a pound without any trouble. This year he dropped five cents on the price about the middle of April, and it was for this reason I had gotten private custom for part of it. About two weeks later he dropped to eighteen cents a pound. This circumstance made me first consider the idea of packing butter.

For years I had been using sugar in my butter, and every body using it wondered why it was that my butter never got strong. Many times people have asked me why it was that my butter never got strong. They would sometimes get enough to last them for several weeks and yet the last of it would be as good as the first had been. So by experience I knew that I might succeed in keeping the butter for a number of weeks without losing it, but I did not know how it would be about keeping it for months right through the heat of the summer. While in this predicament I got to thinking one day. That when I put up pork I always used half as much sugar as I did salt and why should not the same plan work with butter. I will try it, I thought, and try it I did, and with the *best of success.*

WHAT MAKES BUTTER STRONG.

There are a thousand and one things that will make butter strong, but I have embraced them all in the simple rules explained in these pages. Any one folfowing them will surely succeed. It has been said that only one woman in twenty was naturally a butter maker, nevertheless, it is supposed that the other nineteen are intelligent enough to learn how to make good butter. I think I have sufficiently explained about feed, cleanliness, etc. All I have said about them must be followed if you would have your butter good. And if it is not *strictly first class* when put up you will not find it so after being packed for six months. Once more I would say to those intending to pack their butter, observe what I have said about heating your milk pans. The cream jar and butter crocks should also be heated. I heat my cream jar by setting on the stove-hearth, keep turning until it has been heated all round. All butter crocks that have grease fry out of them should be well cleansed again. The fire in this instance, as in many others, proves to be the "All Purifing Element,"—hot water will not do, as it will not penetrate the pores of the earthenware, and the particles of oil and milk will still be there, though we may scald them a dozen times.

Remember as the leaven that leaveneth the whole measure, so the rancid oil or particle of rotten milk will spoil a whole jar of butter.

Feed has much to do with the taste and flavor of butter. It may make butter strong. Indeed, I need not tell those whose cows have ever gotten into an old cabbage field and have eaten frozen or rotten cabbage, or those whose cows delight in a good mess of garlic, how disagreeable milk and butter can be. Indeed, *real garlic* butter can run any one out of the house.

(*Note.* Let me say to those troubled with garlic butter, if you will take this butter, set on the stove, and let it cook slowly until it is clear, when cold the flavor of garlic will have left, and it can then be used for cooking and baking.)

While this butter is strong the cows make it so, and all that the bntter maker can do will not make it fit for the table. But where the milk is good, the butter should also be good.

The best time for packing butter is during the months of May and June.

At this time of the year the nights are still cool, and the young grass gives the butter a delicious flavor which makes it really finer than the butter made at any other time of the year. There is one great advantage connected with packing our butter. We can in this way keep the surplus made during these two months until later on when butter is less plentiful and sometimes is indeed very scarce. At this time we can put our packed butter onto the market and get not only good prices but much credit for it.

PACKING BUTTER.

Now if your butter is just what it ought to be when taken out of the churn, you must work it properly, using fine table salt and granulated sugar as I have said before in the article on salting butter. A teacup of salt and half as much sugar for every six pounds of butter. If the weather is very warm and the butter is at all soft when you work it, it may be necessary to give it a second working in order to get the milk all out. To do this: Take your butter and put in a crock, set on the ground in the cellar or milk house and cover with a plate or inverted crock; let it set for twenty-four hours. It will then under ordinary circumstances be firm. Now turn all back into your butter tray and you will find that you can get the milk out. Do this early in the morning while the air is still cool. Should the weather be stormy and sultry, it will not hurt the butter to set for two or even three days without this second working. After working all the milk out, you can either pack in small crocks or bulk it in jars, whichever suits best. Fill your crocks within an inch and a half from the top, and then fill this space up with a brine made strong enough to bear an egg, using the same proportions of sugar and salt as in salting the butter.

This brine should be boiled and skimmed, then set away to cool before using. When you pack in jars, empty the brine each time you are going to put more in the jar. Press your butter very evenly so there may be no raised places, and when you have it all level put the brine back on. Now one thing more. These crocks and jars must not be covered so as to *exclude the air.* Take a piece of *coarse* cloth and tie them up so as to keep insects and dust out.

I packed about 150 pounds in three months, beginning to pack the first of May. The summer of 1895 proved to be so very dry and pastures were so short that butter was as scarce in August as it was plentiful in May, and all of my packed butter went to my regular customers giving the best satisfaction. I saved the first lot I packed to use myself, and the last of it was as good about Christmas time as it was when put up.

In conclusion, I will say that every

word I have written here is *tried* and *true.* There is no theorizing, but all has been practically applied. I have tried not to repeat the same thing over, and have said what I had to say in as few words as I could to make it plain. Several years ago, I wrote a few articles on butter making. They were so well received that after studying the matter over I came to the conclusion that these remarks might be acceptable to the public. If these lines should find their way into many homes, I hope they will be well received and made welcome.

BERTHA BRIDGES.

Sweet Wine, *November* 12, 1896.

www.ingramcontent.com/pod-product-compliance
Lightning Source LLC
Chambersburg PA
CBHW021440090426
42739CB00009B/1569